HOW TO BUILD AN ELEPHANT

A SYNTHESIS OF PHYSICS, CHEMISTRY AND BIOLOGY FOR ANIMAL LOVERS BIG AND SMALL

MARK SCHREIBER

CANAL HOUSE

to Scott Goldsmith
with whom I built dinosaurs

Illustrated by Kim Lattimer, Greg Sandor and Mark Schreiber

ISBN 978-0-9754664-2-1

Library of Congress Control Number: 2007900378

So you want to build an elephant? Well you've come to the right place. Here's all you'll need:

UP QUARKS: 400,000,000,000,000,000,000,000,000,000

DOWN QUARKS: 200,000,000,000,000,000,000,000,000,000

ELECTRONS: 300,000,000,000,000,000,000,000,000,000

"Wait a minute!" you ask. "What about the cells and atoms? Isn't everything made of atoms?"

Sure, but you want to build your elephant from scratch, right? If not, you could just go to the Elephant Organ Store, buy a pair of elephant lungs, a liver, heart, brain, trunk, etc. and then stop by the Elephant Graveyard, pick up a skeleton and bam, you've got your elephant. But where's the challenge in that?

Here's a question before we begin. What is an elephant chiefly made of? Water, you say? After all, you've heard that people are mostly water. And it's true, in a sense. But in a deeper sense, people and elephants are mostly made of...nothing!

Now don't try telling that to a baobab tree that's just been obliterated by a stampeding herd. But if you were a subatomic particle called a neutrino, this is what an elephant would look like:

There are no zoos in the neutrino world, no safaris. If you were a neutrino you could pass through a whole family of elephants without a scratch.

"Cool!" you say. "So where can I pick up some of these neutrinos and how much do they cost?" You're on a budget, after all. Not to worry, though. You would need neutrinos if you were building a star, but not for an elephant. In fact, we should warn you, there are all kinds of things people are going to try to sell you that you don't need.

For starters, when you walk into Quarks R Us, I guarantee you the salesman is going to look at your list of up quarks and down quarks and say, "But there are six quarks in all: up, down, top, bottom, strange and charm, not to mention their antiquarks: antiup, antidown, antitop, antibottom, antistrange, and anticharm. So twelve quarks altogether. That'll be $99.50."

Well hold your ground. It's true there are six quarks in all, and six antiquarks. However these other quarks don't exist very long in living things but rapidly decay into up and down quarks.

The same holds true at the Electron Depot. Some overzealous clerk who doesn't know the Big Bang from a black hole in the wall is going to point to aisles of neutrinos, which we've already discussed, as well as muons, tau particles and a hermetically sealed annex of antimatter and insist you need them all. Tell him you're only building an elephant, not a universe.

A reminder: When you buy your quarks make sure they include the gluons.

Now some quark stores have been known to charge the unsuspecting customer extra for gluons. But gluons are part of the deal. Quarks send them back and forth to other quarks as a means of hanging together, much as far-flung family members send messages on the telephone to hang together. Actually, it's nothing like that at all, but you get the point.

Gluons, despite their small size and extreme anorexia are the strongest force in the universe. "How can this be?" you ask. "Surely gravity is the strongest force? Gravity makes the earth shake each time an elephant takes a step. Elephants could fly if it weren't for gravity. What do gluons do that concerns our elephant?"

Well, despite appearances, gravity is actually the weakest force in the universe. Elephants and planets may feel it, but just try asking a quark what gravity is. Gravity can't touch a quark. But if gluons didn't bind quarks together, all our protons—which we'll get to soon—would repel each other instead of staying in the atom's nucleus and our poor elephant would drift apart like 200,000,000,000,000,000,000,000,000 balloons on a windy day.

By the way, it's no use shopping around for quarks. Quarks R Us is the only game in town. Because the strong force binds quarks so tightly together, there are no other sources of free quarks. We would have to go back to the beginning of the universe, just after the Big Bang, to find enough free quarks flying around to build our elephant.

Electrons, like quarks, also come with a free particle that doesn't weigh anything and also carries a force. This particle is the photon (not to be confused with the proton). It is pure energy and travels at the speed of light. The force it carries is electromagnetism. Photons give us electricity, light, heat, X-rays, radio and TV. So now you know whom to blame.

A DAY IN THE LIFE OF AN ELECTRON

Be happy you're not an elementary particle. It's very mundane. You might exist for millions and millions of years doing pretty much the same thing, or your existence might be so short that scientists call it "virtual." As in virtual photons, which are the photons exchanged between nearby electrons.

As for electrons, they travel around the nucleus of the atom. They may change energy states, or be pulled toward a neighboring atom, and rarely they may get knocked out of the atom, but for the most part they stay put. That means that if you take your suitcase from New York to Tokyo, the electrons in the atoms in the suitcase also take the trip. But if the electrons are in gold atoms far under the earth, they aren't going anywhere unless there's an earthquake or a lucky miner discovers them.

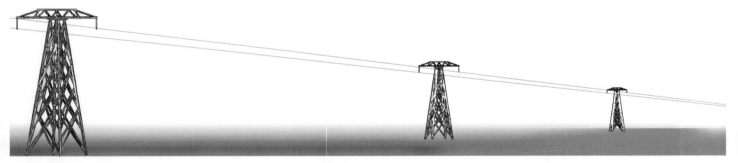

But what about electricity? Don't electrons in a wire travel across the country in seconds flat? Unfortunately, electricity isn't the glamour profession some electrons believe. They think they're lucky because they're in those big wires strung up on a power line. They think they are going to see the world. But this kind of electricity—non-battery electricity—is called AC. That means alternating current. That means the electron goes a short distance, bumps the electron in front, bounces back, gets bumped by the electron behind, and so on.

But at the end of the day it's still in Kansas.

A DAY IN THE LIFE OF A PHOTON

Unfortunately, it's even more boring being a photon than an electron. And you may have a severe identity crisis.

At least most electrons survive forever. A photon only survives until it hits something, which may be almost immediately, in the case of a virtual photon, or never, as in the case of the light from a star that no one ever sees.

But let's take a photon from the sun. It leaves the sun's surface at the speed of light. Photons don't have gas pedals or brakes. It's speed of light all the way.

About seven minutes later it reaches the earth, where it crashes into a tree on the Serengeti Plain. The tree wins. Our photon's day is over.

Now what about an elephant looking at that same tree from his watering hole? The reason an elephant can see the tree in the first place is that light reflects off the tree and hits the elephant's eye. And of course light means photons.

But the photon that came from the sun is not the same photon that hit the elephant's eye. It is not like throwing a ball against the tree, where the same ball bounces back.

The photon that hit the tree is absorbed. Don't try looking for it. Because it is an elementary particle, it can't break into something smaller. It has simply disappeared.

But the laws of energy normally require a photon to be reflected from the tree. And it is this new photon that hits the elephant's eye.

Where did the new photon come from? The Twilight Zone? Actually, science is much stranger than science fiction.

You're ready to build an atom. Put together two up quarks and one down quark and you've built a proton. A proton is a subatomic particle with a positive electric charge.

Put together two down quarks and one up quark and you've built a neutron. A neutron is a subatomic particle with no electric charge.

Reminder: You aren't smashing the quarks together like peanut butter and jelly. Remember that most everything is empty space. In fact, a proton or neutron takes up about a thousand times more space than a quark. The quarks are placed just close enough together so that they can exchange gluons, which transmit the strong force and bind them.

Take your proton. This serves as the smallest possible atomic nucleus. (Larger atoms have more than one proton, and one or more neutrons in their nuclei.)

Then toss a negatively-charged electron into the mix and, congratulations, you've built your first atom!

You may be wondering why the electron doesn't fall into the nucleus with the proton. That's because electrons aren't affected by the strong force. The strong force, although it's the strongest force in the universe, only works at the very short distances between quarks.

But what about charge? Electrons are negative and protons are positive. Don't opposites attract? They do. But the electron's speed—it usually travels about one percent the speed of light—keeps it from falling into the nucleus.

Now I'm sure you've seen drawings or models of atoms with balls representing electrons orbiting other balls representing protons and neutrons. Forget these, they're wrong.

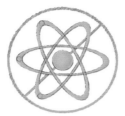

Remember, an atom is mostly empty space. If the nucleus were the size of a stitch on a baseball, the electron would be flying in the parking lot outside the stadium.

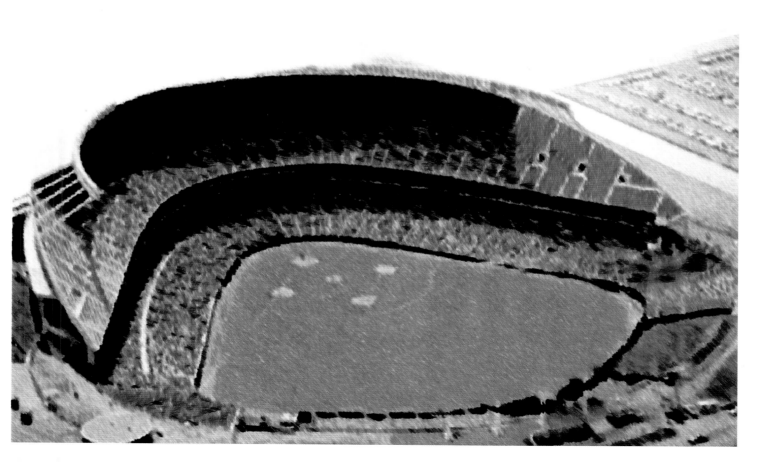

Furthermore, these particles are not the same size. Protons and neutrons are about 1,000 times heavier than electrons.

Furthermore, there's a lot more going on inside an atom than those models would lead you to believe. It is a simplification to say there are only three quarks in a proton or neutron. Actually there are an indefinite number of quarks and antiquarks flying around inside protons and neutrons. But the antiquarks cancel out the quarks, except for two up quarks and one down quark for a proton, and two down quarks and one up quark for a neutron.

And remember our gluons and photons? Gluons are being exchanged all the time within the atomic nucleus, keeping it from flying apart. And photons are constantly being ejected and captured by electrons and protons. In fact all of our senses, everything we touch, see, taste, smell and hear is the result of those photons coming and going. So now you know whom to thank.

When an electron absorbs a photon it gains energy and soars into a higher energy level (farther from the nucleus). When an electron emits a photon it loses energy and sinks closer to the nucleus.

We can determine the energy level of an electron but not its precise location. If we could see an atom we would see an electron cloud, like one of the rings of Saturn. The electron would be somewhere in the cloud.

Electrons also spin. They can spin clockwise or they can spin anti-clockwise.

No more than one electron with the same spin can occupy the same cloud in a particular energy level in an atom at the same time.

When an electron jumps from one energy level to another it does not pass Go. It does not collect two hundred dollars. One moment it is in one energy level, the next moment it is in another. This tricky move is called a quantum jump.

18

Finally, and here's the tricky part, electrons, photons and gluons aren't just particles, they're also waves. No, they don't make waves, like drops of water make waves. And no, they aren't particles sometimes and waves sometimes. Take a deep breath. Each photon, electron and gluon is both a particle and a wave at the same time! It's rough on illustrators, but that's the universe for you.

Now you may have heard about String Theory and multiple universes and extra dimensions of space.

Fortunately, our elephant only lives in three dimensions of space. Besides, there are competing string theories and none of them has yet been confirmed by experiment.

String theorists imagine the universe composed of very tiny strings that form different particles by vibrating at different frequencies. So that an up quark would be a one-dimensional string vibrating at a certain rate, a down quark would be a string vibrating at a different rate, and an electron would be a string vibrating at a still different rate.

The important point for our task is that whether quarks and electrons are particles or strings, they cannot be divided further.

The atom you've built, the simplest atom possible, with no neutrons and only one proton and one electron, is an atom of hydrogen. And that's not a bad start because hydrogen is the most common atom in an elephant, as well as in the entire universe. Three-fourths of the identified matter in the universe is hydrogen. And as you well know, two atoms of hydrogen and one atom of oxygen make one molecule of H_2O, or water. And H_2O is the most common molecule in an elephant, but we'll get to that later.

There are about 118 kinds of atoms in the universe. You make different atoms by adding protons, neutrons and electrons.

For example:

A carbon atom has 6 protons, 6 neutrons and 6 electrons.

An oxygen atom has 8 protons, 8 neutrons and 8 electrons.

A potassium atom has 19 protons, 20 neutrons and 19 electrons.

(Atoms also contain photons and gluons, in no set amount. They simply appear and disappear.)

Not all types of atoms are present in an elephant. For example, an elephant has no uranium atoms—so we hope.

Most of the atoms in an elephant are:

carbon

hydrogen

nitrogen

oxygen

sodium

The next step is to put atoms together into molecules.

On the subatomic level, even with all the unstable particles and antiparticles, there are still only a few hundred elementary particles.

And we've just seen that there are only about 118 atoms.

But there are countless molecules in the universe, especially on earth, where scientists create new ones every day.

Atoms are discrete. They do not combine with other atoms to form new atoms.

But molecules can combine with other molecules to form new molecules.

Molecules stay together due to the electromagnetic force, most commonly when adjacent atoms share one or more outermost electrons. A shared outer electron is as much attracted to the positively charged proton in the other atom as it is to the proton in its own nucleus.

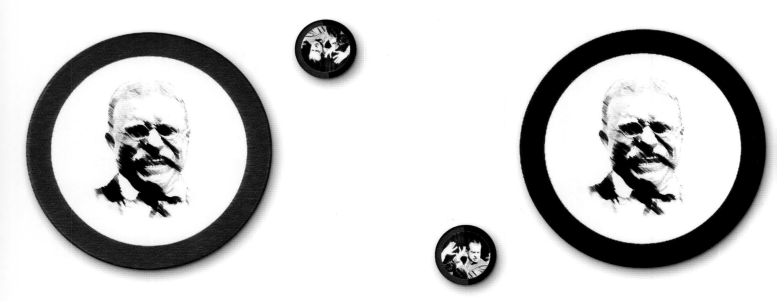

Of course, outer electrons from different atoms also slightly repel each other, as both are negatively charged. These attractions and repulsions allow atoms to combine into molecules without losing their identity as atoms.

Molecules can be very simple, such as a water molecule with its two atoms of hydrogen and one of oxygen. Or they can be very large, such as a plastic molecule, which can contain billions of smaller molecules. (Hundreds of billions of atoms.)

Molecules come in various shapes as well. Some bind their atoms like spokes on a wheel, others like links in a chain. Some are relatively flat, like tiles. Others extend more equally in three dimensions, like bricks.

The bad news is that an elephant contains millions of different kinds of molecules. Some of the most plentiful are:

water

sugar

amino acids

proteins

phosphate

Amino acids are simple molecules made from oxygen, hydrogen and nitrogen atoms, arranged in a cluster.

There are about twenty different kinds of amino acids in an elephant.

These twenty amino acid molecules combine to make proteins. Proteins are very large molecules and very diverse.

There are about 100,000 different proteins in an elephant.

Proteins give structure to cells, which we'll discuss soon, and carry out many important functions in the body, such as the hemoglobin protein, which transports oxygen atoms through the blood.

Enzymes are also proteins. Enzymes carry out chemical reactions without being changed by the reaction. Like Uncle Fred burning hot dogs on the grill.

About half the dry weight of a elephant's cell is protein.

The good news is that you don't have to make all of these protein molecules yourself. They will be faithfully copied by other molecules, which you can think of as workers and engineers.

These worker and engineer molecules follow a blueprint. The blueprint is also a molecule—DNA.

DNA is an extremely long but very narrow molecule. If stretched out it would reach over five feet in length, but only one-trillionth of an inch wide. It is made up of billions of very simple molecules.

Since DNA is so important to building an elephant we will go in to some detail here.

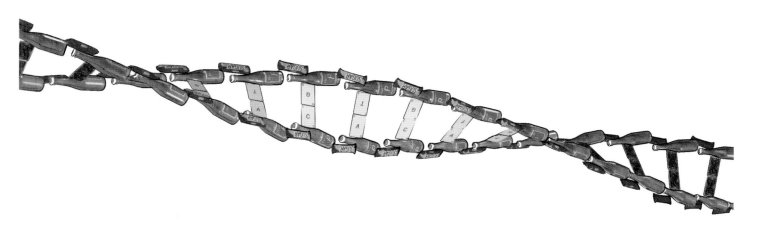

The molecules you will need to build a DNA molecule are:

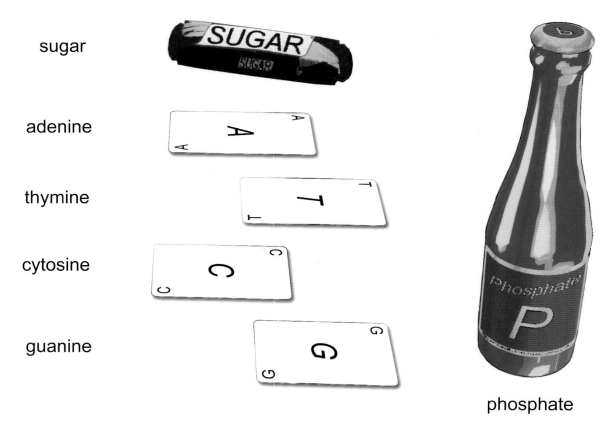

sugar

adenine

thymine

cytosine

guanine

phosphate

These molecules are made from oxygen, hydrogen, carbon, nitrogen, and phosphorus atoms.

You've probably heard that DNA looks like a double helix.

The helices, which are like the sides of a spiral staircase in a grand casino, are made from phosphate and sugar molecules—good sources of energy for the game of life about to be played.

The four other molecules are dealt out in pairs.
Adenine, or A, is always dealt with thymine, T. (Think "at.")

Cytosine, C, is always dealt with, G. (Think "Catherine the Great.")

Three pairs can be cashed in for an amino acid. Depending on the cards, there are twenty kinds of amino acid that can be won.

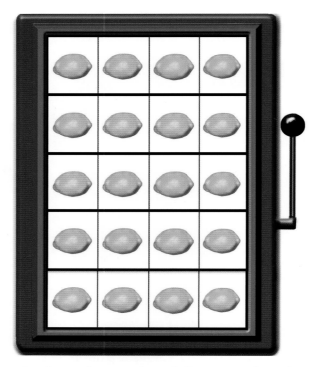

But the smart play is a hand of cards made of thousands of pairs. This is a gene and can be cashed in for a protein. And these proteins determine certain characteristics and functions, such as size, shape, eye color, skin color, and so on.

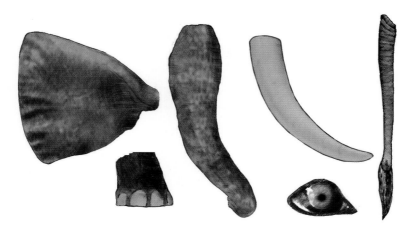

And that's what DNA is. A series of genes that is continually being cashed in for proteins.

There are about 80,000 genes in the DNA of an elephant.

The DNA molecules tells a story. The story of who you physically are. Or who your elephant is. Change a few important genes and you have a hummingbird instead of an elephant. So be careful here.

This may seem complicated, but don't be discouraged. You only have to build DNA once. After that it can copy itself abundantly.

So, we now know quarks and electrons make atoms, atoms combine to make molecules, and molecules combine to make the universe—everything from oceans and mountains to bicycles and alligators.

Most molecules combine to make inorganic matter—things that aren't alive.

But if DNA gets thrown into the mix, then the DNA molecule combines with other molecules to form cells, and where there are cells, there is life.

A cell is a watery chemical laboratory.

A cell can grow and reproduce itself through division. In other words, it multiplies by dividing!

Because a cell is alive, it can also die.

Atoms can lose or gain electrons, and sometimes they decay.

Molecules can be reduced to other molecules or ruptured into atoms.

But atoms and molecules do not live or die.

An amoeba has only a single cell.

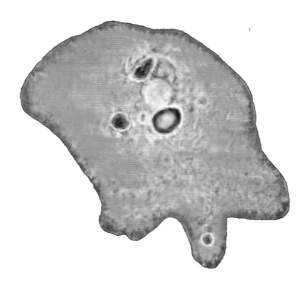

But an elephant contains trillions of cells. And because they are alive, cells are constantly dying and being born as the elephant goes about its day.

A cell contains thousands or millions of molecules.

There are many types of cells and they vary in size and shape. Some look like baseball bats, some like baseballs, some like wads of chewing tobacco spat in the dugout. Like molecules, they vary in size, but most are microscopic.

Some cells in an elephant are nerve cells, or neurons. Others are blood cells, muscle cells, sperm cells, egg cells and so on. These cells perform different functions, but they all share certain characteristics.

Each cell has a thin membrane made of proteins and fats. (The fats are called lipids.)

The membrane cleverly lets some substances into the cell but not others. (Sometimes they are not so clever and let in viruses.)

The main body of the cell is called the cytoplasm.

If the membrane is the wall of the laboratory, the cytoplasm is the main room. But remember, cells are watery and there is no floor or ceiling.

And just as a laboratory contains different machines to perform different tasks, so does the cytoplasm.

Mitochondria are structures made from many molecules which serve as the engine of the cell, transforming glucose (or sugar) and oxygen into energy. They do this using enzymes to initiate chemical reactions.

There may be thousands of mitochondria in the cytoplasm of one cell.

The cytoplasm also contains thousands of ribosomes.

Ribosomes are tiny factories that make proteins. A ribosome is made up mainly of about a thousand protein molecules. Think of a ribosome as a brick building that makes more bricks.

Just as a laboratory has a safe where its secret formulas are protected, so the cell has a secure nucleus, separated from the cytoplasm by its own membrane.

The cell's secret formula is its DNA.

You're probably wondering how the DNA molecule, which is over five feet long if stretched out, can fit into a cell.

You may also have wondered how flight attendants can fit a week's worth of clothes into a carry-on when you need two large suitcases. Remember, DNA is only one-trillionth of an inch wide. It folds up like an army cot with three billion hinges.

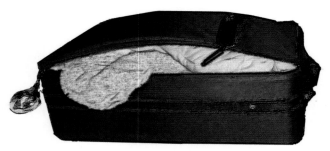

Actually, DNA is not one unbroken molecule that can be stretched to over five feet. It is actually 54 separate double-helix molecules—like the sections of the Sunday *New York Times*.

Each section gets wrapped around some protein molecules like thread around a spool and then gets covered by other protein molecules for protection.

This whole package is called a chromosome.

Different animals have different numbers of chromosomes. All mammals have two sets of chromosomes, one inherited from the mother and one from the father.

A human has 24 chromosomes from the mother and 24 from the father. (Two of these chromosomes, one from each parent—X from the mother, and X or Y from the father—determine gender and do not contain DNA.)

An elephant has 56 chromosomes, 28 from its mother and 28 from its father. But only 54 contain DNA.

So the elephant cell's nucleus has two sets of 27 molecules of DNA, one from the father and one from the mother, stored in chromosomes, plus two chromosomes that determine its sex.

The father's set of DNA and the mother's set of DNA both have the same genes. For instance both sets have a gene for eye color. If the father's gene is for black and the mother's is for green, one must prevail. The winner is called the dominant gene. The loser is called the recessive gene.

You've undoubtedly heard of cloning and may be wondering if you are cloning an elephant. I assure you you are not. You are building an elephant, from the quarks and electrons up, a much harder task.

To clone an elephant all you have to do is take an egg cell from a female elephant, remove the nucleus, take the chromosomes from any cell of your donor elephant, inject these into the egg, put the egg into an elephant uterus and let it replicate. Hardly worth a manual, is it?

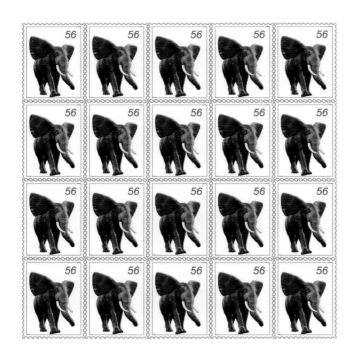

A cloned elephant will not have 28 chromosomes from its mother and 28 from its father. All its chromosomes, and DNA, will be the same as the donor elephant.

By the way, the animal's complete DNA can be referred to as that animal's genome.

(Not to be confused with a gene, which, as you remember, is a blueprint for a protein or group of proteins that determine a hereditary characteristic.)

Just as a factory doesn't order everything in a supplier's catalog but only the parts it needs, so the ribosome only orders the proteins it needs.

Along with proteins, the ribosome also contains many molecules of RNA. RNA is similar to DNA, except it only has one strand, or helix, and is not as long.

RNA travels from the ribosome to the nucleus.

A specialized enzyme already in the nucleus unzips a DNA molecule. Not all the way, but just for that part that needs to be copied. But don't think these enzymes possess the intelligence of librarians, who can find the right book among thousands. Rather it is as if every book in a library had its own librarian.

As we know, the DNA base molecules A,T, C and G can be cashed in for amino acid molecules. But they can also be used to signal where the many genes along the DNA double helix start and stop.

The RNA from the ribosome is specialized to go to one particular gene. It finds the place to start and uses spare A, T, C and G molecules floating around in the nucleus to copy the gene. When it reaches the stop signal it breaks away with its partial copy of DNA and returns to the ribosome.

In the ribosome, amino acid molecules are floating around in quantity. Another kind of RNA, called transfer RNA, fits the amino acids to the corresponding bases on the partial copy of DNA. For example, the bases AT AT CG correspond to the amino acid lysine.

So pretty soon we have a series of amino acid molecules attached to the base molecules on the partial copy of DNA.

The process is completed by yet another kind of RNA, called ribosomal RNA, which binds the amino acids together in a long chain called—a protein!

So DNA molecules are partly unzipped to make specific proteins.

DNA molecules are fully unzipped to make complete copies of all the DNA in the elephant. In other words, to copy the elephant's genome.

There are two reasons for DNA to copy itself in this way. The first is to make new cells, such as hair cells or muscle cells or blood cells, and so on. This is necessary because cells are always dying.

"How then," you might ask, "can there be such different cells as muscle cells and tusk cells and kidney cells when they all contain the same DNA?"

The answer is that the genes in DNA can be turned on and off. So in a muscle cell the genes necessary for a muscle cell are turned on, while those for tusks, kidneys, etc. are turned off.

When DNA copies itself to make a cell for health reasons, for instance to replace dying skin cells, or to make more white blood T-cells to fight invaders, the cell divides by mitosis.

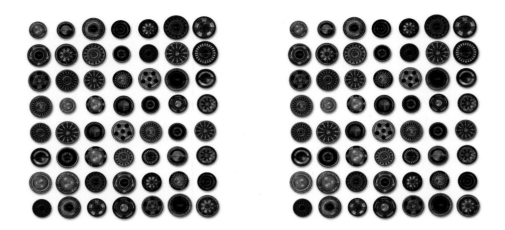

But DNA also copies itself to make a new elephant, by making an egg or sperm cell. In this case the cell divides by meiosis.

For a cell to divide by mitosis, all the DNA first has to be unzipped by enzymes. Remember, there are 56 chromosomes in an elephant's cell, 28 from the mother and 28 from the father. But two of these chromosomes determine sex and do not contain DNA.

To make a specific protein, RNA only copied the part of the DNA molecule it needed. But to make a new cell, RNA needs to copy the entire length of all 54 DNA molecules.

In meiosis the new cells—which are sperm or egg cells—contain only half the original chromosomes.

Sperm and egg cells still contain all the genes necessary to make an elephant, but instead of having two sets of genes—one from the mother and one from the father—meiosis reduces them to one set, randomly mixed.

In other words, the sex cell might have the eye color gene from the mother and the hair color gene from the father. When the egg cell with its single set of genes is fertilized by a sperm cell with its single set of genes, the baby elephant cells again have two full sets of genes, but not identical to its parents' genes.

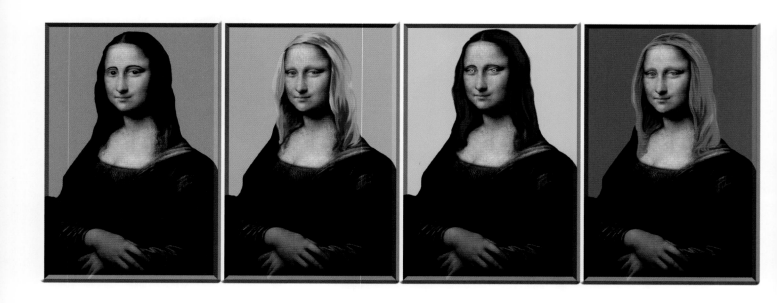

The random mixing of genes in meiosis accounts for diversity and evolution.

Since mitosis and meiosis sound similar—job security for biologists—remember that T cells are made by mitosis.

Cells are constantly working, absorbing nutrients, making proteins, carrying out their particular functions, such as destroying invaders or growing bone and muscle. And of course, except for nerve cells, which normally don't divide, they are constantly copying themselves.

All this activity going on in the cell is called metabolism.

Let's take a look at your work so far:

You'll need a microscope:

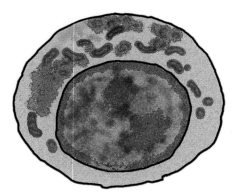

Granted, a cell in a petri dish may not look like much but, believe me, you're more than halfway home.

The problem with building inorganic structures like houses is that you have to lay every brick yourself. And if you want to build a second house you have to start from scratch. Living structures build themselves because the DNA molecule can replicate itself, and the genes in the DNA molecule tell the cells how to build proteins.

So now you can sit back and supervise because you've got 80,000 genes working for you, and they don't belong to a union!

A NOTE ABOUT VIRUSES:

A virus is not a cell and most scientists don't consider it living, although there is some debate.

A virus is a molecule of either DNA or RNA wrapped in a protein shell. Some also have a lipid (fatty) shell. That's it.

Viruses wreak havoc after entering the body through the nose or mouth or skin. They attach themselves to the membrane of a cell. The cell believes the virus is friendly and allows it to penetrate into the cytoplasm.

Inside the cell the virus uses the cell's own enzymes to uncoat its protein shell. Then the virus's DNA or RNA swims into the nucleus.

The cell foolishly copies the virus DNA or RNA as if it were its own and keeps copying until the cell bursts, sending the virus DNA or RNA to attack other cells.

Usually white blood cells, which are large cells responsible for destroying invaders, catch on to the plot and eliminate the virus.

But sometimes the virus causes too much damage and the elephant dies.

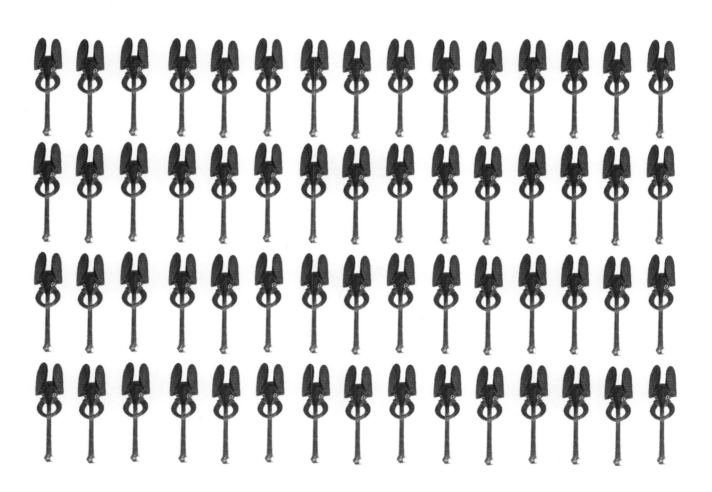

A NOTE ABOUT BACTERIA:

If you put all living things on your bathroom scale, bacteria would be the dominant life form on earth!

Granted, they are only one cell large and need to be seen with a microscope, but there are so many of them that combined they outweigh all the elephants or all the humans on the planet!

They have also been around for much longer than any other life form and live in more places, from the eyelashes of a human to the ocean floor.

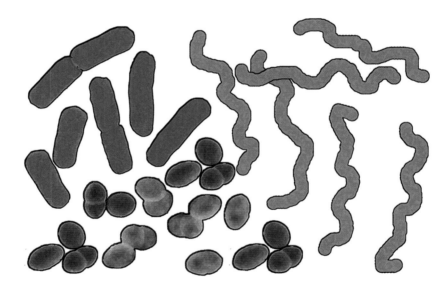

Bacteria are much smaller than elephant cells, and they actually outnumber elephant cells in the elephant's own body. But most bacteria are helpful or harmless. Unlike an elephant's cells they don't have a nucleus. Under the right conditions they divide every twenty minutes or so. Of course, they also die very quickly.

Like-minded cells band together to form tissues. There are many kinds of tissues in our elephant, such as nerve tissue, muscle tissue, connective tissue and so on.

Tissues in turn combine to form organs.

Organs can contain billions of cells.

The most important organs are the brain and the heart. The largest organ is the skin. Yes, the skin is considered an organ. So is the trunk.

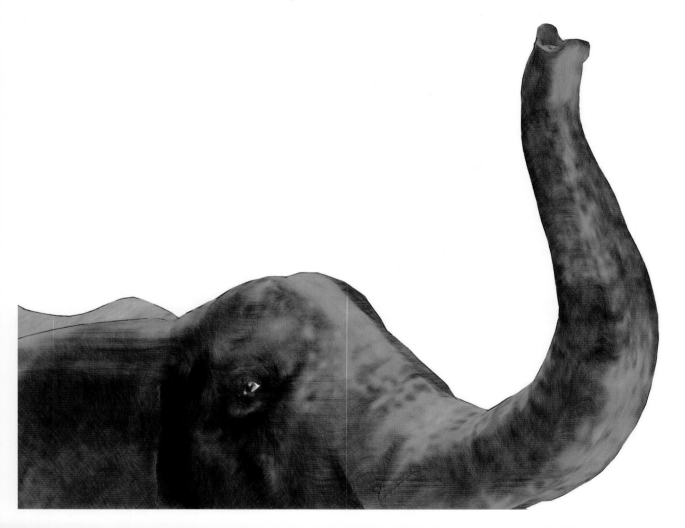

Organs work together in systems to performs certain functions.

Some of the most important systems are the respiratory system, the circulatory system, the digestive system, the nervous system, and the skeletal system.

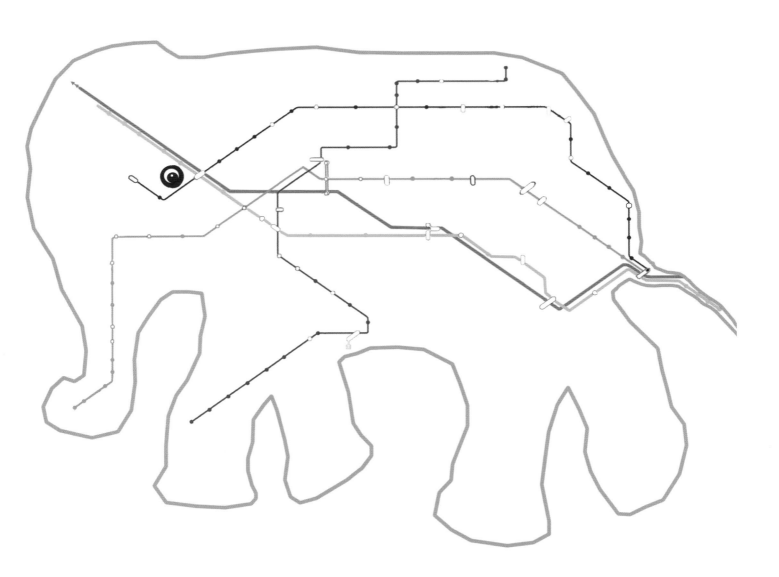

58

The respiratory system consists mainly of the lungs, trunk and trachea.

By inhaling air the respiratory system brings oxygen atoms into the body. Oxygen is needed by the cells to perform chemical reactions and grow.

But our air is not pure oxygen. There are other atoms and molecules as well, mainly nitrogen atoms.

The inhaled air goes to the lungs, where it is filtered.

The air which is not oxygen atoms is then exhaled in the form of carbon dioxide and water vapor molecules.

The circulatory system consists mainly of the heart, veins, arteries, capillaries and blood cells.

The oxygen that is filtered through the lungs is carted off by red blood cells, which act like the post office, delivering oxygen atoms to every address in the body through tiny vessels called capillaries. Every cell in the body lies next to a capillary, like mailboxes along a road.

Of course, just as packages sometimes get lost by the post office, there are sometimes problems in delivering oxygen. One circulatory disease, anemia, is caused by the failure of red blood cells to deliver enough oxygen.

Blood cells are pumped through the body by the heart. Besides red blood cells, there are white blood cells, which defend the body, and platelet cells, which rush to injuries to form clots.

The heart, after the brain, is the most important organ in the body because oxygen is so important. If cells do not receive oxygen they begin to die. Nerve cells in the brain begin to die after a few minutes.

It seems there are many ways for an elephant to die. It can simply grow so old that its cells stop dividing. (This happens, by the way, because over time the DNA in a particular cell grows shorter and shorter until that cell will no longer be able to be copied.)

Our elephant could be shot by a poacher. It could die from starvation, or a virus, or by inhaling smoke in a fire, and on and on. But basically death comes down to one thing—starving the brain of oxygen, so that the nerve cells die.

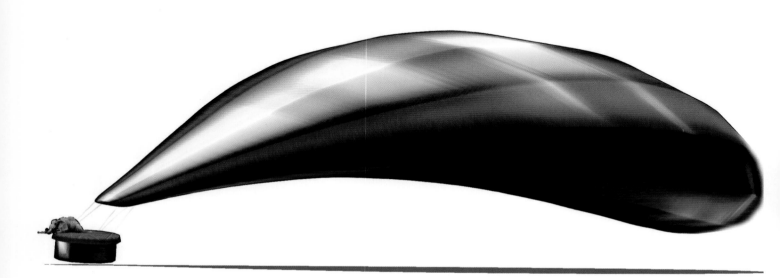

There is another circulatory system, called the lymphatic system, which circulates—you guessed it—lymph.

Lymph makes up the liquid part of blood, called plasma. Lymph also contains lymphocytes, which are a type of white cell important in fighting infections. (Pus is lymph on a bad day.)

The lymphatic system does not have a heart. Instead lymph is circulated by normal muscle contractions.

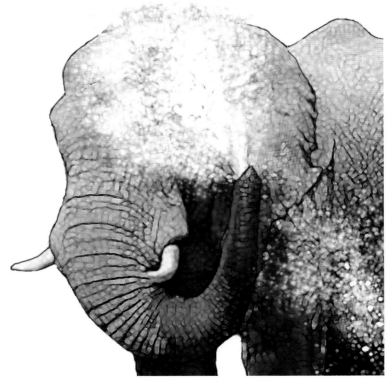

By the way, since we're speaking of fluids, most cells in the body are not directly connected to each other. Nor do they float in empty space. They are surrounded by interstitial fluid which, of course, is mostly water.

Back to the bloodstream:

About a million red blood cells are produced every minute, but they only live a few months. A red blood cell is made of hundreds of molecules of the protein hemoglobin.

After the red blood cells collect oxygen from the lungs they travel to the heart, which pumps them through arteries.

The red blood cells drop off oxygen to cells and pick up carbon dioxide, which is a waste product of the cells' metabolism. That's why saving the rain forests is so important. If it weren't for plants, we'd eventually use up all the oxygen in the air, replacing it with carbon dioxide. But fortunately one organism's trash is another organism's treasure—plants take in carbon dioxide and expel oxygen.

The red blood cells carry the carbon dioxide back through veins. (Veins appear blue in humans not because they lack oxygen but because they are far enough under the skin that only the blue wavelengths of light—or blue photons—are reflected.)

Finally, the red blood cells return through the heart and back to the lungs, where the carbon dioxide is exhaled, along with water vapor.

Whew!

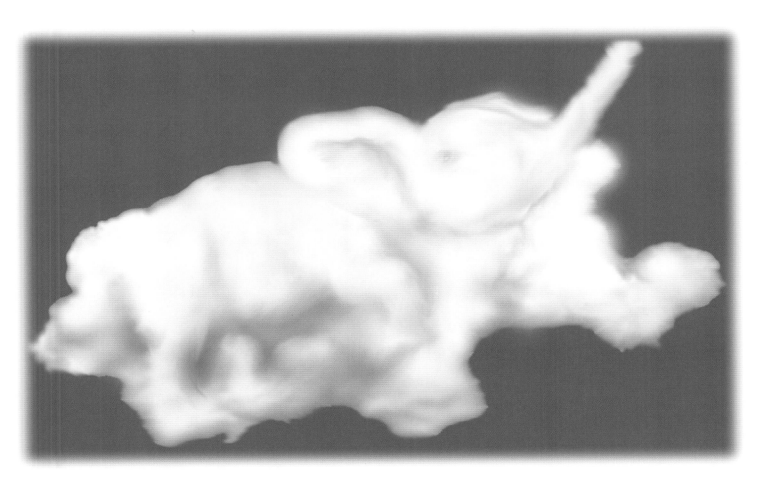

The digestive system is a little more relaxed but a lot messier. Nutrients in the form of food and drink—grasses and water for our elephant, and sometimes peanuts—travel from the mouth, through the tube called the esophagus and into the stomach.

Enzymes in the stomach break the food down into a soupy mixture, which proceeds to the small intestine.

Meanwhile the cells in the liver are busy making bile, which is then stored in the gall bladder. (Bile used to be called "gall.")

And the pancreas is making pancreatic juice (what else)?

Bile and pancreatic juice are enzymes, which enter the small intestine and break down the food molecules into amino acids, two sugar molecules (glucose and fructose), and fatty acid molecules.

These smaller molecules can be dissolved in water and therefore are able to enter a cell's membrane.

These amino acids, sugars and fatty acids are distributed to all the cells in the body through the blood and lymphatic vessels, where they provide the cells with energy to grow and divide.

The food that cannot be broken down into these small dissolvable molecules proceeds to the large intestine, or colon. Water is removed and the rest is passed out through the anus.

The feces of an elephant are extremely large, not only because the elephant is so large, but because it doesn't efficiently digest food—only about 44% of what an elephant eats is used by its cells. The stomach of a human is very acidic and can break down meat and even bubble gum. The stomach of a shark is even more acidic and can break down bones and perhaps even license plates if given enough time.

But the stomach of an elephant is not very acidic. That is why it's a vegetarian. It can't even break down most of the grasses it eats. That is why if you look at an elephant's feces—not a required activity, but if you've nothing better to do—you will see entire blades of undigested grass.

Liquid waste, of course, exits the body as urine.

Three of the most basic liquids in mammals are:

the liquid in blood, which is called plasma, or lymph
the liquid inside cells
and interstitial liquid, which is the liquid between cells.

These liquids have to maintain a certain balance in how much salt, acid and other atoms and molecules they contain. The kidneys act as a filter, removing the excess.

This is why an elephant cannot drink salt water. Its bodily fluids would become imbalanced by drinking so much salt, and in removing it the kidneys would have to pour out more water than the elephant drank.

The nervous system consists mainly of the brain, the spinal cord, and nerve cells, or neurons.

The nervous system is responsible for connecting the elephant to the world by allowing it to see, feel, taste, hear and see.

If elephants have long memories it is because of the nervous system.

The nervous system is also responsible for connecting the elephant to itself, by transmitting chemical messages from the brain to the cells and responding to messages from the cells to the brain.

About half the elephant's neurons are in its brain.

But not all signals have to reach the brain. Reflexive movements travel from a sensory neuron on the skin to the spinal cord, and back to a motor neuron that causes a contraction in the muscle.

Neurons are long and thin and vary greatly in size. Some neurons in the spinal cord are several feet long.

At one end are branches that receive signals. At the other end are branches that send signals. There can be thousands of these branches, like the branches on a baobab tree.

Neurons generally do not divide like other cells. But they can form new connections with other neurons. This is what happens when the elephant learns. New connections are formed between the branches of neurons in its brain.

A normal atom has no electric charge because it has as many negative electrons as positive protons. But if it loses one or more electrons it becomes positively charged. If it gains electrons it becomes negatively charged. Charged atoms are called ions.

The interstitial fluid between cells is like weak salt water, so it's no surprise there are a lot of sodium ions floating around. There are also a lot of potassium and calcium ions.

These ions are also present in neurons. But some proteins on the membrane act like gates, keeping the sodium level lower inside than outside. When a neuron is stimulated these gates briefly open. Positive sodium ions rush in from the interstitial fluid, adding to the sodium and potassium ions already inside the neuron, generating an electric current.

This triggers protein gates at the other end of the cell to let in calcium ions, which cause molecules called neurotransmitters to leap across a very tiny gap, called a synapse, to an adjacent neuron. This action sparks an electric charge on that neuron, and on and on.

The excess sodium, potassium and calcium ions get pushed back out. And the neurotransmitters return like bowling balls after a strike, ready for the next impulse.

So the next time you have a negative thought, blame these three positive ions.

Of course, all these organs and systems would fall to the ground if they didn't have a skeleton to support them. Our skeleton has 206 bones. But a male African elephant has 351 bones. They are assembled as follows:

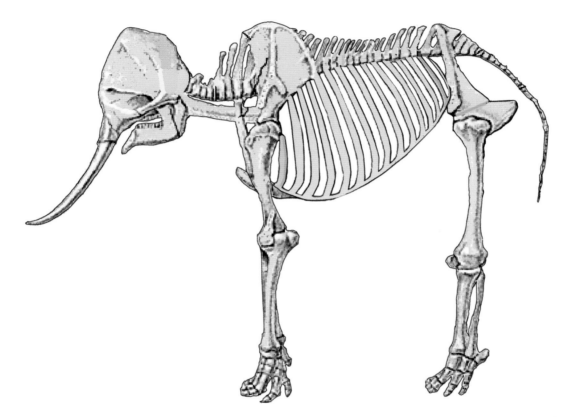

The bones are made from water and protein molecules, along with molecules of minerals such as calcium.

There are even bone cells, which either destroy old bone or create new bone. If it weren't for bone cells, broken bones could not heal.

And within the bone is a spongy substance called marrow, which produces blood cells.

Now you are ready for the final assembly. Here is how your elephant should look with all the bones and organs in place, except for the skin.

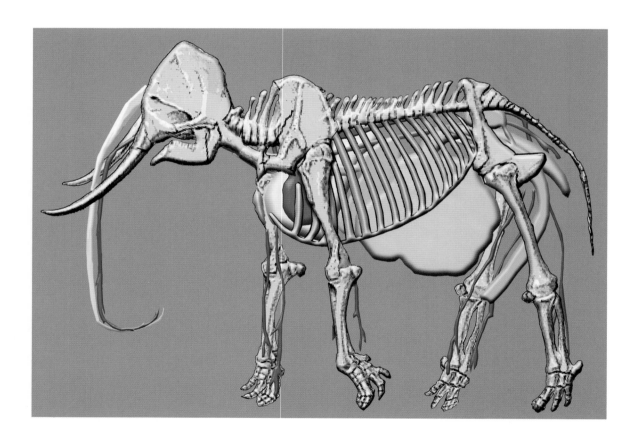

And with the skin:

Congratulations! You've built an elephant. If he isn't run over by an elephantine SUV he could live for seventy years!

Now all you have to do is feed him.

FRIENDS

Theoretical physicist **Greg Kilcup** served unknowingly as the inspiration for this book, for it was an impromptu three-hour private tutorial on particle physics in his kitchen one night that led me to consider the relationship between quarks, electrons, atoms, molecules, cells and elephants. His Grand Tour of the Ivy League began in early childhood when his father was a graduate student at Harvard. After attending high school in Seattle, Greg entered Yale at age seventeen with an ambitious triple-major in Philosophy, Mathematics and Physics. He received his Ph.D. at Harvard under Howard Georgi, and completed postdocs at Cornell and Brown. He was recruited to The Ohio State University by Nobel Prize winner Ken Wilson to work in a new field called Lattice Guage Theory, which Wilson had created to study the properties of quarks. Greg has received numerous prizes, including The Ohio State University Alumni Award for Distinguished Teaching, which is based on evaluations by students and colleagues and given to only ten teachers each year out of a faculty of nearly three thousand. Greg's passions include chess, scuba diving, pool, the harpsicord and supercomputers, and he excels in all of them to a degree that would make him insufferable were it not for his disarming, generous nature.

Siegfried Kra is an Associate Professor of Cardiology at Yale. He has written numerous books for the general reader, as well as textbooks, and has published scores of articles, including studies in *Lancet* and the *New England Journal of Medicine*. An avid tennis player, his passions also include creative writing, French cuisine, opera, and growing orchids. And we must add medicine, for at age seventy-four he still treats patients, lectures on hypertension and preventive cardiac disease and indefatigably leads third-year medical students through the wards of Yale-New Haven Hospital and the Hospital of Saint Raphael.

The rain forest couldn't have a better advocate than **Samara Karla Nasciemento Jäehnke**. I first met Samara at the Ariau Towers resort in the Amazon, where she had been hired by the owner to install an exhibit of local animals and insects. She received her Masters in Biochemistry from the University of Manaus, for work on the uses of medicinal plants against snake venom, and her Ph.D. in Physical Chemistry at the University of Mainz for research on the complexation of metals in medicinal plants. Her great passion is life itself. She even has something nice to say about mosquitoes. "The mosquito—not the jaguar—is the true guardian of the forest."

Although I graduated high school at fifteen and never went to college, I consider myself a lifelong student. I believe we sometimes learn the most when we don't even realize we're learning, when it's fun. Not only are the sciences considered daunting subjects, but their increasing specialization makes it difficult to relate concepts and structures from one field to another. My own knowledge deficit despite years of reading general science books, coupled with a desire to bridge the gap between disciplines and between science and art, education and pleasure, led to my writing this book. Other books of mine include *Dreams of the Solo Trapeze: Offstage With the Cirque du Soleil*, and the novels *Princes in Exile*, *Pebble Beach*, *Starcrossed* and *Carnelian*.